NUTRITION ANALYSIS POCKET GUIDE

Allyn and Bacon
Boston · London · Toronto · Sydney · Tokyo · Singapore

A Simon & Schuster Company
Needham Heights, Massachusetts 02194

ISBN 0-205-19958-5

Printed in the United States of America

10 9 8 7 6 5 4 3 2 1   00 99 98 97 96 95

CONTENTS

PREFACE

The U.S. Department of Health and Human Services reports that over one-half of all deaths in the United States are associated with health problems linked to poor nutrition. The leading killers in the United States—cardiovascular disease, cancer, obesity, and diabetes—are all diseases linked to dietary excess in calories, sugar, fat, and sodium. But don't despair! Through diet analysis and modification, it is possible to prevent many of these nutrition-related diseases. The first step is awareness. We have developed this Nutrition Analysis Pocket Guide as a handy resource to help you increase your awareness and take charge of your health by improving your diet.

To help you get a handle on your overall eating habits, we have included diet analysis worksheets to record your nutrient intake over a 3-day period. The nutritional content of common foods and nutritional content of fast foods tables provide the key information you will need to complete these worksheets. At the end of the 3-day period, you will be able to compare your average nutrient intake to the recommended dietary allowances. Based on this comparison, you will be able to identify where you need to make modifications in your diet. We have even provided additional worksheets for a 3-month and a 6-month check-up to track your progress. After you have completed the diet analysis worksheets, you may wish to use the sample diet forms included to design a new diet that meets the recommended dietary allowances.

In addition to improving your diet, taking charge of your health involves increasing the amount of energy you expend each day through physical activity. The key is to find a balance between the amount of energy you take in and the amount of energy you expend. To help you determine your caloric balance, we've included the energy costs of various activities. Our goal is to help you achieve a healthy lifestyle. Take the first step and use this pocket guide to begin taking charge of your health!

PART ONE

Nutritional Content of Common Foods*

This food composition table has been prepared for Allyn and Bacon and is copyrighted by DINE Systems, Inc., the developer and publisher of the DINE System family of nutrition software for personal computers. The values in this food composition table were derived from the USDA Nutrient Data Base for Standard Reference Release 10 and nutrient composition information from over 300 food companies. Nutrient values used for each food were determined by collapsing similar foods into one food, using the median nutrient values. In the food composition table, foods are listed within the following eight groups. The eight groups are, in order: fruits, vegetables, beverages, alcoholic beverages, grains, dairy, fats/sweets/other, and protein foods. Further information can be obtained from:

DINE Systems, Inc.
586 N. French Road
Amherst, NY 14228
(716) 688-2492
FAX (716) 688-2502

NUTRITIONAL CONTENT OF COMMON FOODS

NOTE: Carbohydrate value does not include Added Sugar value

	AMOUNT	CAL	PROT (gm)	T FAT (gm)	S FAT (gm)	CARB (gm)	SUG (gm)	FBR (gm)	CHOL (mg)
GRAINS									
Bagel	1 bagel	175	7	10.22	0.22	32.5	2.25	1.5	0
Barley-cooked	1/2 cup	84	2.25	0.11	0	18.5	0	2.5	0
Biscuit	1 biscuit	100	2	3.78	1.22	13.5	0	0.5	2
Bread or roll-wheat	1 slice, 1 roll	65	2	0.89	0.22	10	2	1.4	0
Bread or roll-white	1 slice, 1 roll	70	2.75	1.11	0.33	11	2	0.5	0
Bread-mixed grain	1 slice	65	2	0.89	0.22	9.75	2.25	1.4	0
Bread-oatmeal	1 slice	90	4	1.78	0.33	10.5	4	1.5	0
Bread-pita, wheat	1 pita	145	5.5	0.89	0.22	28.5	0.5	4.1	0
Bread-pita, white	1 pita	160	6	1	0.22	30	1	1.1	0
Bread-raisin	1 slice	70	2	1	0.33	8.25	3	0.9	0
Bread-rye, pumpernickel	1 slice	80	2.75	0.89	0.11	13	1	1.1	0
Bread-wheat	1 slice	82	3	1	0.33	11.38	2	1.8	0
Bread-wheat, diet	1 slice	40	2.5	0.56	0.11	4.25	2	2	0
Bread-white	1 slice	67	2	1	0.22	11	1	0.7	0
Bread-white, diet	1 slice	40	3	0	0	6	1	2	0
Bread-whole wheat	1 slice	69	2.63	1.22	0.17	13	1.38	2	0
Breadsticks	1/3 large, 2 small	36	1.25	0.78	0.22	6	0.5	0.6	0
Bulgur	1/2 cup	113	4.25	0.22	0	23.5	0	3	0
Cake-nonfat	1 piece	140	3	0	0	32	15.5	1.2	0
Cereal-bran, fiber	1/3 cup	62	2	0.78	0.22	7.75	2.75	3.8	0
Cereal-granola	1/4 cup	130	3	4.89	0.89	13.5	4.5	1.7	0
Cereal-granola, fat free	1/4 cup	90	2	0	0	21	0.25	2.5	0
Cereal-oat flakes	1 cup	182	5	1.78	0.44	24.75	4.5	3.2	0
Cereal-other, cold	1 cup	110	2	0.22	0.11	19.5	3	0.6	0

Cereal-sweetened	1 cup	120	1.25	1	0.44	13	13	0.5	0
Cereal-whole grain	3/4 cup	105	3	0.78	0.22	19	0	2.3	0
Cheese ravioli w/sauce	1 cup, 6 pieces	314	13	10.83	6.11	45.13	0.25	3.4	30
Cornbread, hushpuppies	1 square, 3 hushpuppies	166	3	5.33	1.78	25.75	0	1.6	42
Couscous	2/3 cup	120	4	0	0	26	0	4.3	0
Cracker sandwiches	2 large, 5 small	70	1.75	3.11	1.11	7.75	0.75	0.2	1
Crackers	5 crackers	60	1	2	0.44	8.75	0.75	0.4	0
Crackers-butter type	5 large, 10 small	70	1	3.56	1.11	8	0.25	0.2	1
Crackers-crispbread	1 large, 2 small	40	1	0.89	0.33	8.5	0	1.6	0
Crackers-lowfat	2 large, 5 small	60	2.25	2.33	0.33	10	0	0.6	0
Crackers-wheat	4 crackers	70	1	2.22	0.44	8	0.5	1.5	0
Croissants	1 croissant	310	7	19	11.22	27	6	2	0
Croutons	4 tablespoons	46	1	1.78	0.44	6.25	0.25	0.3	0
English muffin	1 muffin	130	5	0.89	0.33	25	1.5	1.6	0
French toast	1 slice	151	4.25	7.33	1.89	16.25	3.5	0.5	48
Fried rice	1/2 cup	165	4	5.56	1	18.25	0.75	0.6	14
Grits	1/2 cup	73	1.5	0.11	0	16	0	0.3	0
Lasagna-meat	1 serving	350	27.5	23.11	6	31	1.5	2.6	73
Lasagna-vegetable	1 serving	315	22	12	5.89	27	0.5	4.6	40
Macaroni	1/2 cup	95	3.25	0.33	0.11	19	0	0.9	0
Macaroni and cheese	1/2 cup	191	6	8.56	2.67	22.75	0	1.4	18
Macaroni-whole wheat	1 cup	202	8	1.11	0.22	39	0	7.4	0
Matzo or melba toast	1 matzo, 5 melba	100	3.25	0.56	0.11	22.25	0	0.9	0
Meat ravioli with sauce	1 cup, 6 pieces	324	13	8.33	2.89	46.75	0.75	3.1	20
Muffins	1 muffin	240	4.5	8.5	1.67	40	9	2	24.5
Muffins-fat free	1 muffin	165	3	0.17	0	38	10.63	1	0
Noodles-chow mein	1/2 cup	130	3	6.44	1.11	14.5	0	0.5	2
Noodles-egg	1/2 cup	106.5	3.75	1.17	0.22	19.88	0	0.9	26.5
Noodles-egg, macaroni	1 cup	190	7	0.56	0.11	37	0	1.1	0
Oatmeal-flavored	1 packet	140	4	2.22	0.44	18	8.25	2.4	0

NUTRITIONAL CONTENT OF COMMON FOODS (continued)

NOTE: Carbohydrate value does not include Added Sugar value

	AMOUNT	CAL	PROT (gm)	T FAT (gm)	S FAT (gm)	CARB (gm)	SUG (gm)	FBR (gm)	CHOL (mg)
Oatmeal-plain	2/3 cup cooked	109	4	1.78	0.33	18.25	0	3.3	0
Pancakes	1 pancake	73	2	1.22	0.44	11	1.25	0.5	7
Pasta w/parmesan cheese	1/2 cup	252	6.5	14.67	6.33	22.5	0	1.1	38
Rice cake	1 large cake	35	0.75	0	0	7.5	0	0.3	0
Rice-brown	1/2 cup	115	2.5	0.67	0.22	25	0	1.8	0
Rice-long grain & wild, mix	1/2 cup	137	3	2	0.56	23	2	0.9	0
Rice-seasoned	1/2 cup	150	3.5	3.78	1.67	23	1	1	5
Rice-white	1/2 cup	92	2	0	0	20.5	0	0.5	0
Roll-hamburger/hotdog, wheat	1 roll	114	4	1.11	0.22	21.25	0.75	2.5	0
Roll-hamburger/hotdog, white	1 roll	138	3.25	2.44	0.67	22.5	2	1.4	0
Roll-hoagie, sub	1 roll	400	11	7.11	1.78	68.5	4	1.8	0
Roll-wheat	1 roll	77	2.5	1.78	0.44	11.5	1.5	1.1	0
Roll-white	1 roll	105	3.25	1.67	0.44	16.25	1.75	0.6	0
Salad-pasta	1/2 cup	250	4	16	3.33	20.75	0	0.7	28
Spaghetti	1 cup	200	7.5	1	0.22	40.5	0	1.1	0
Spaghetti w/meatballs	1 cup	307	12	10.56	3.89	37	0	2.7	34
Spaghetti-whole wheat	1 cup	200	9	1.11	0.22	39.5	0	5.8	0
Stuffing	1/2 cup	210	4.5	12.67	3	17.5	2.5	0.5	22
Taboule	1/2 cup	170	3	8.67	1.33	20	0	1.6	0
Taco shell	1 shell	50	0	2	1.11	8	0	0.2	0
Tortilla	1 tortilla	65	2	1	0.11	12	0	0.9	0
Waffles	1 waffle	110	3	4.33	0.89	13	1.75	0.8	3

VEGETABLES

Artichokes	1/2 cup, 1/2 vegetable	36	2	0.11	0	7.25	0	2.6	0
Asparagus	6 spears, 1/2 cup	24	1.75	0.33	0.11	3.75	0	1.5	0
Asparagus-canned	1/2 cup	21	1.5	0.33	0.11	2.75	0	1.9	0
Bamboo shoots	1/4 cup	4	0.25	0	0	0.5	0	0.3	0
Bamboo shoots-canned	1/4 cup	6	0.5	0.11	0	1	0	0.6	0
Bean sprouts	1/2 cup	16	1.25	0	0	2.75	0	2.2	0
Bean sprouts-canned	1/2 cup	8	0.75	0	0	1.25	0	2.3	0
Beets-canned	1/2 cup	36	0.75	0	0	8	0	2.9	0
Beets-pickled	1/2 cup	82	0.75	0	0	8.5	11.25	2.8	0
Beets-raw, cooked	1/2 cup	31	0.75	0	0	6.75	0	2.5	0
Bok choy-chinese cabbage	1/2 cup	5	0.25	0	0	0.75	0	0.6	0
Broccoli-cooked	1/2 cup, 2 spear	24	1.75	0.11	0	4.25	0	2.5	0
Broccoli-raw	1/2 cup	12	0.75	0.11	0	2	0	1.5	0
Brussels sprouts	1/2 cup, 4 sprouts	32	1.5	0.33	0.11	5.75	0	2.4	0
Cabbage-raw or cooked	1/4 cup cooked, 1/2 cup raw	9	0.25	0	0	1.75	0	1	0
Carrots-canned	1/2 cup	17	0.25	0.11	0	3.75	0	2.7	0
Carrots-raw or cooked	1/2 cup, 6 sticks	26	0.5	0	0	5.75	0	2.2	0
Cauliflower-raw or cooked	1/2 cup	15	0.75	0.11	0	2.5	0	1.3	0
Celery-raw or cooked	1/2 cup, 6 sticks	10	0.25	0	0	2.25	0	1.4	0
Chef salad	1 cup	25	2.5	0.17	0	4	0	2	0
Coleslaw	1/2 cup	154	0.5	14.44	2.67	4.25	1	1.5	7
Corn	1 ear, 1/2 cup	80	1.75	0.33	0	17.5	0	4.6	0
Corn-canned	1/2 cup	83	1.5	0.44	0.11	13	5.5	4.7	0
Cucumber	1/2 cup	7	0.25	0	0	1.5	0	0.7	0
Eggplant	1/2 cup	13	0.25	0	0	3	0	1	0

NUTRITIONAL CONTENT OF COMMON FOODS (continued)

NOTE: Carbohydrate value does not include Added Sugar value

	AMOUNT	CAL	PROT (gm)	T FAT (gm)	S FAT (gm)	CARB (gm)	SUG (gm)	FBR (gm)	CHOL (mg)
French fries	1 regular, 1-1/4 cup	241	3	12.72	3.89	29.5	0	2.8	0
Fried vegetables/onions	1/2 cup, 6 onion rings	180	2	10.78	2.67	15.5	0	1	0
Green beans-canned	1/2 cup	13	0.75	0	0	2.5	0	1.5	0
Green beans-raw or cooked	1/2 cup	20	1	0	0	4	0	1.8	0
Greens-collard & mustard	1/2 cup	16	1.75	0.22	0	3.75	0	1.4	0
Greens-mustard, turnip, cooked	1/2 cup	15	1	0.11	0	2.5	0	1.4	0
Greens-mustard, turnip, raw	1/2 cup	7	0.5	0	0	1.5	0	0.9	0
Greens-turnip, canned	1/2 cup	17	1	0.22	0.11	2.5	0	2.2	0
Kale-raw or cooked	1/2 cup	20	0.75	0.11	0	3	0	1.7	0
Lettuce-endive	1/2 cup	4	0.25	0	0	0.75	0	0.7	0
Lettuce-iceberg	1 cup	7	0.5	0.11	0	1.25	0	0.6	0
Mixed vegetables-canned	1/2 cup	39	1.25	0.11	0	8	0	3.5	0
Mixed vegetables-frozen	1/2 cup	22	0.75	0	0	4.5	0	2	0
Mushrooms-canned	1/2 cup	19	1	0.11	0	3.25	0	1.1	0
Mushrooms-fresh, cooked	1/2 cup	25	1.75	0.11	0	4.25	0	1.5	0
Mushrooms-raw	1/2 cup	9	0.5	0	0	1.5	0	0.6	0
Okra	1/2 cup	26	1	0.11	0	5.5	0	2.1	0
Onions	1/2 cup	29	0.5	0.11	0	6.25	0	1.4	0
Parsnips	1/2 cup, 1/2 vegetable	64	0.75	0.11	0	14.75	0	3.2	0

Peas-green	1/2 cup	63	3.75	0.11	0	11.5	0	3.5	0
Peas-green, canned	1/2 cup	59	3.25	0.11	0	8	2.75	5.3	0
Peas-snowpeas	1/2 cup	35	2.25	0.11	0	5.75	0	3.4	0
Peppers-hot	2 tablespoons	8	0.25	0	0	1.5	0	0.2	0
Peppers-sweet, green	1/2 cup, 1/2 vegetable	12	0.25	0.11	0	2.25	0	0.6	0
Peppers-sweet, red	1/2 cup	12	0.25	0.11	0	2.25	0	0.8	0
Potato skins-cheese, bacon	2 halves	302	11	15.89	7.44	27.5	0	1.4	34
Potato-baked/boiled	1/2 baked, 1/2 cup	73	1.5	0.11	0	16.75	0	1.4	0
Potatoes-mashed	1/2 cup	118	1.5	4.56	1.44	17	0	1.8	4
Radishes	2 radishes	2	0	0	0	0.25	0	0.1	0
Romaine lettuce	1 cup	9	1	0.11	0	1.25	0	1.3	0
Rutabaga	1/2 cup	35	0.75	0.11	0	7.5	0	2.1	0
Salad-potato	1/2 cup	153	3	7.78	1.89	15	1.5	1.9	47
Salad-three bean	1/2 cup	80	2	0	0	16.25	1.75	5	0
Salsa	1/4 cup	32	0.75	0	0	4	0	0.9	0
Sauerkraut	1/2 cup	22	0.75	0.11	0	4.5	0	4.1	0
Vegetables (continued)									
Soup-vegetable	1 cup	81	2	1.56	0.44	11.25	2.25	0.5	2
Spaghetti sauce	1/2 cup	118	2	5	0.89	12.75	1.5	3.3	0
Spaghetti sauce w/ meat	1/2 cup	80	2	3.11	0.67	12.5	2	3.3	2
Spinach-canned	1/2 cup	25	1.75	0.33	0.11	3.25	0	3.9	0
Spinach-fresh, cooked	1/2 cup	24	1.75	0.11	0	3.75	0	3	0
Spinach-raw	1/2 cup	6	0.5	0	0	0.75	0	1.1	0
Squash-summer	1/2 cup	18	0.5	0.11	0	3.75	0	1.5	0
Squash-winter	1/2 cup	41	0.75	0.11	0	9.5	0	3	0
Squash-zucchini-fresh, cooked	1 cup raw, 1/2 cup cooked	18	0.75	0	0	3.5	0	1.4	0

NUTRITIONAL CONTENT OF COMMON FOODS (continued)

NOTE: Carbohydrate value does not include Added Sugar value

	AMOUNT	CAL	PROT (gm)	T FAT (gm)	S FAT (gm)	CARB (gm)	SUG (gm)	FBR (gm)	CHOL (mg)
Sweet potato	1/2 cup	98	1.25	0	0	23	0	4	0
Sweet potato-candied	1/2 cup	190	1	0	0	26.5	20	4.4	0
Tomatoes-canned or stewed	1/2 cup	34	0.75	0.11	0	6.5	0.25	2.4	0
Tomatoes-raw	1/2 cup, 4 slices	17	0.5	0.11	0	3.5	0	1.3	0
Vegetable juice	3/4 cup	35	1	0	0	8	0	1	0
Waterchestnuts-canned	1/2 cup	34	0.25	0	0	8.25	0	1.5	0
Waterchestnuts-raw	1/2 cup	66	0.75	0	0	15.75	0	1.4	0
Watercress-raw	1/2 cup	2	0.25	0	0	0.25	0	0.2	0
Wax beans	1/2 cup	18	1	0.11	0	4.25	0	1.5	0
Yams	1/2 cup	69	1.25	0.11	0	16.75	0	1.8	0
FRUITS									
Apple cider	3/4 cup	94	0	0.11	0	17	6	0.1	0
Apples-sweetened	1/2 cup	68	0	0.11	0	12.5	3.25	2	0
Apples-unsweetened	1 fruit, 1/2 cup	77	0.25	0.44	0.11	20	0	2.1	0
Applesauce-sweetened	1/2 cup	97	0.25	0	0	12	11	1.8	0
Applesauce-unsweetened	1/2 cup	53	0.25	0	0	12.5	0	2.5	0
Apricots-sweetened	1 fruit	65	0.5	0	0	3.75	10.75	1.1	0
Apricots-unsweetened	1 fruit, 2 canned	17	0.5	0.11	0	4	0	0.8	0
Avocados	1/2 fruit, 1/2 cup	166	2	14.22	2.56	7.25	0	2.7	0
Banana	1 fruit, 1/2 cup	105	1	0.33	0.22	24	0	2.3	0
Blueberries-sweetened	1/2 cup	103	0.5	0.11	0	8.5	16.25	2.3	0
Blueberries-unsweetened	1/2 cup	41	0.5	0.22	0	9	0	1.7	0
Cherries-sweetened	1/2 cup	106	0.75	0	0	12.25	13	0.9	0
Cherries-unsweetened	10 fruits, 1/2 cup	44	0.75	0	0	10	0	0.9	0
Dates	5 fruits, 1/4 cup	[illegible]	[illegible]	[illegible]	[illegible]	[illegible]	[illegible]	[illegible]	[illegible]

Dried fruit	1/4 cup, 8 pieces	92	1	0	0	21.75	0	2.4	0
Figs-sweetened	2 fruit	45	0.25	0	0	7	3.75	1.3	0
Figs-unsweetened	4 fruit, 1/2 cup	144	1.63	0.61	0.11	37	0	6.4	0
Fruit cocktail	1/2 cup	56	0.5	0.11	0	14.5	2.5	1.4	0
Fruit cocktail-sweetened	1/2 cup	83	0.5	0	0	8.75	11	1.4	0
Fruit cocktail-unsweetened	1/2 cup	50	0.5	0	0	11.5	0	1.4	0
Grapefruit-sweetened	1/2 cup	76	0.5	0	0	10	8	0.5	0
Grapefruit-unsweetened	1/2 fruit, 1/2 cup	39	0.5	0	0	9	0	0.7	0
Grapes-sweetened	1/2 cup	94	0.5	0	0	10.5	12	0.5	0
Grapes-unsweetened	20 fruits, 1/2 cup	48	0.5	0	0	11.5	0	0.5	0
Guava	1 fruit	45	0.5	0.33	0.11	9.5	0	5	0
Juice-unsweetened	3/4 cup	90	0.5	0	0	22.5	0	0.2	0
Kiwi fruit	1 fruit	46	0.75	0.33	0.11	10	0	2.1	0
Mango	1/2 fruit, 1/2 cup	61	0.5	0.11	0	14.25	0	1.8	0
Melon	1/2 cup	30	0.5	0	0	7	0	0.7	0
Nectarines	1 fruit	68	1	0.56	0	14.5	0	3.3	0
Olives	3 olives	15	0.25	1.56	0.22	0.25	0	0.3	0
Orange	1 fruit, 1/2 cup	63	1	0	0	14	0	3.9	0
Papaya	1/2 fruit, 1/2 cup	56	0.75	0	0	12.75	0	1.8	0
Peaches-sweetened	1/2 cup	94	0.5	0	0	5.25	17	1.2	0
Peaches-unsweetened	1 fruit, 1/2 cup	44	0.5	0	0	10.75	0	1.3	0
Pears-sweetened	2 halves	103	0.5	0	0	6	18.5	3.2	0
Pears-unsweetened	1 fruit, 1/2 cup	98	0.75	0.67	0	25	0	4.3	0
Pineapple-sweetened	2 slices, 1/2 cup	93	0.5	0	0	7.5	15	0.9	0
Pineapple-unsweetened	2 slice, 1/2 cup	70	0.25	0	0	17.5	0	0.9	0
Plums-sweetened	2 plums	67	0.5	0.11	0	8	7.75	0.7	0
Plums-unsweetened	1 raw, 2 canned	37	0.5	0	0	8.25	0	0.9	0
Prunes-cooked	1/2 cup, 7 fruits	136	1.25	0.33	0	32	0	4.9	0
Prunes-dried	1/2 cup	209	2	0.44	0	49.25	0	6.8	0
Pumpkin-canned	1/2 cup	41	0.75	0.11	0.11	8.75	0	3.5	0
Raisins	1/4 cup	109	1	0	0	26	0	2.5	0

NUTRITIONAL CONTENT OF COMMON FOODS (continued)

NOTE: Carbohydrate value does not include Added Sugar value

	AMOUNT	CAL	PROT (gm)	T FAT (gm)	S FAT (gm)	CARB (gm)	SUG (gm)	FBR (gm)	CHOL (mg)
Raspberries-sweetened	1/2 cup	117	0.75	0.11	0	12	18.25	4.2	0
Raspberries-unsweetened	1/2 cup	30	0.5	0.22	0	6.5	0	3	0
Strawberries-sweetened	1/2 cup	100	0.5	0.11	0	9	15	2	0
Strawberries-unsweetened	1/2 cup	24	0.5	0.22	0	5.5	0	1.6	0
Tangerines-sweetened	1/2 cup	76	0.5	0	0	10	8.5	0.9	0
Tangerines-unsweetened	1 fruit, 1/2 cup	43	0.5	0	0	9.75	0	0.9	0
Watermelon	1/2 cup	25	0.5	0.22	0.22	5	0	0.3	0
DAIRY									
Buttermilk	1 cup	99	8.75	2	1.33	11.25	0	0	9
Cheese spread	2 tablespoon	81	3.5	6.56	4.33	2	0	1	89
Cheese-American	1 ounce, 1 slice	106	6.75	8.22	5.44	0.5	0	0	27
Cheese-cheddar	1 ounce, 1 slice	113	7.5	8.67	5.89	0.5	0	0	30
Cheese-cottage	1/2 cup	113	14.5	4.78	3.11	3	0	0	17
Cheese-cottage, nonfat	1/2 cup	90	14	0	0	7	0	0	10
Cheese-cottage, lowfat	1/2 cup	96	15	1.22	0.78	4	0	0	5
Cheese-mozzarella	1 ounce, 1 slice	80	6	5.67	3.67	0.5	0	0	22
Cheese-mozzarella, light	1 ounce	72	7.25	4.11	2.78	0.75	0	0	16
Cheese-nonfat	1 ounce, 1 slice	40	8	0	0	1	0	0	5
Cheese-parmesan/romano	1 tablespoon	20	2	1.33	0.89	0.25	0	0	4
Cheese-provolone	1 ounce, 1 slice	100	7.75	7	4.78	0.5	0	0	20
Cheese-reduced fat	1 ounce, 1 slice	80	8	5	3	1	0	0	20
Cheese-ricotta	1/2 cup	216	15	14.89	10	3.75	0	0	63
Cheese-ricotta, part skim	1/2 cup	166	14.5	8.67	5.67	6.25	0	0	37

Hot cocoa prepared w/milk	1 cup	218	8	9	5.67	13.5	11.25	3	33
Ice cream	1/2 cup	148	2.5	7.44	4.67	4.5	11.75	0.2	30
Ice milk	1/2 cup	110	3	2.78	1.89	10	8	0.3	8
Lowfat chocolate milk	1 cup	175	8.5	3.78	2.33	12.25	16	1.3	12
Meal replacement drinks	1 cup	200	14	1	0.44	36	17	4	5
Milk-chocolate	1 cup	208	8.5	7.89	5.11	10.5	14.5	1.1	30
Milk-lowfat	1 cup	112	8.75	3.56	2.22	11.25	0	0	14
Milk-skim	1 cup	86	9	0.44	0.33	11.5	0	0	4
Milk-whole	1 cup	150	8.5	7.67	5	11	0	0	33
Tofutti	1/2 cup	150	2.5	6.67	1.11	9	11	1.5	0
Yogurt-frozen	1/2 cup	100	3	1.78	1.11	8	10.25	0.1	7
Yogurt-lowfat w/fruit	1 container	240	9	3	2	27	16	0.3	10
Yogurt-nonfat w/fruit	1 container	95	3.5	0	0	8	12	0	0
Yogurt-plain, lowfat	1 container	142	11.25	3.67	2.44	15.75	0	0	15
Yogurt-plain, nonfat	1 container	110	11	0.22	0.22	16	0	0	4
Yogurt-plain, whole milk	1 container	145	8.75	6.89	4.56	11.5	0	0	32
Yogurt-w/fruit, artificial sweetener	1 container	90	7	0.67	0.44	14	0	0.5	5
PROTEIN FOODS									
Bacon substitute	1 strip	52	3	4.11	1.56	0	0	0	13
Beans-baked	1/2 cup	140	6	1.67	0.67	15	7.5	6	8
Beans-black	1/2 cup	113	6.5	0.33	0.11	20.75	0	4.4	0
Beans-kidney, pinto	1/2 cup	115	6.5	0.33	0.11	22.25	0	4.5	0
Beans-kidney, pinto, canned"	1/2 cup	104	5.75	0.22	0	19.5	0	6.1	0
Beans-lima	1/2 cup	94	5.5	0.22	0.11	17.5	0	4.6	0
Beans-lima, canned	1/2 cup	93	4.75	0.22	0.11	17.5	0	5.8	0
Beans-navy, chickpeas	1/2 cup	132	6.75	1	0.22	23.75	0	4.8	0

NUTRITIONAL CONTENT OF COMMON FOODS (continued)

NOTE: Carbohydrate value does not include Added Sugar value

	AMOUNT	CAL	PROT (gm)	T FAT (gm)	S FAT (gm)	CARB (gm)	SUG (gm)	FBR (gm)	CHOL (mg)
Beans-navy, chickpeas, canned	1/2 cup	146	7	0.78	0.11	27.5	0	5	0
Beans-white, canned	1/2 cup	153	8.25	0.22	0.11	29.25	0	5	0
Beans-white, split peas	1/2 cup	125	7	0.22	0.11	23	0	5.3	0
Beef stew	1 cup	207	15.25	9	4.22	16.5	0.5	2.5	53
Beef-corned	3 ounces	182	18.25	12.11	5.22	0	0.25	0	65
Beef-mixed dish	1 cup	310	19.25	13.56	5.89	23.5	1.25	2.1	68
Biscuit w/egg, meat, cheese	1 biscuit	489	18.75	31.22	9.67	29	4	0.8	347
Bologna sandwich	1 sandwich	311	11	18	6.56	22	3.75	1.7	32
Broadbeans-fava	1/2 cup	93	5.5	0.22	0	17	0	4.4	0
Broadbeans-fava, canned	1/2 cup	91	6	0.11	0	16.25	0	4.5	0
Burrito	1 burrito	213	8	7.22	3.56	29	0	3.2	33
Caviar	1 tablespoon	40	4.25	2.11	0.78	0.5	0	0	94
Cheeseburger (large) & roll	1 sandwich	711	32	43.33	16.78	33	4	1	113
Cheeseburger (lowfat) & roll	1 sandwich	370	24	14	5	35	3.5	1.6	75
Cheeseburger (small) & roll	1 sandwich	461	29	27.56	13.67	25.25	3	0.8	95
Chicken breast sandwich	1 sandwich	509	26	26.89	4.78	34.75	1.75	1.7	83
Chicken fingers/nuggets	4 fingers, 6 nuggets	275	15.75	14	3.11	15.25	0	0.4	49
Chicken salad	1/2 cup	179	14.75	12.22	2.89	0.75	0.75	0.3	118
Chicken wings	10 wings	617	39	46.56	13.56	15	1	0.4	198
Chicken w/skin	3 ounces	189	22.25	9.22	2.56	0	0	0	70
Chicken w/out skin	3 ounces	147	24.5	3.89	1.11	0	0	0	72
Chicken-fried, no skin	4 ounces	107	19.25	4.22	1.33	0.25	0	0	50

Chicken-fried, w/skin	3 ounces	155	15.75	8.44	2.44	4.75	0	0.2	52
Chicken-mixed dish	1 cup	365	15.25	17.78	5.56	13.5	0	1	103
Chickpeas	1/2 cup	138	6.5	1.67	0.11	24.75	0	4.8	0
Chili con carne	1 cup	286	15.75	12.44	5.78	28.5	0	6.5	43
Chili-vegetarian	1 cup	240	18	12	1.78	13	2	16.4	0
Chimichanga	1 chimichanga	425	18.5	17.11	8.33	41.25	0	5.2	30
Chop suey	1 cup	300	26	16	4.33	13	0	1.5	68
Chow mein-beef or chicken	3/4 cup	65	6.5	1.44	0.56	5.25	0.75	1.4	26
Clams, oyster, shrimp-fried	4 pieces	103	5.25	6.11	1.11	6	0	0.1	23
Clams, oysters, shrimp	1/2 cup, 3 ounces	71	12.25	1.22	0.33	2.5	0	0	62
Coconut-shredded	2 tablespoon	44	0.25	3	2.67	2.25	2.25	0.4	0
Crabmeat	3 ounces	86	12.5	1	0.22	4.5	0	0	26
Egg salad	1/2 cup	267	11	22.89	5.78	1	3	0.3	418
Egg-boiled, poached	1 egg	79	6.5	5.56	2.11	0.5	0	0	274
Egg-fried, scrambled	1 egg	89	6.25	6.78	3	1	0	0	281
Egg-omelet	1 omelet (3 eggs)	382.5	24.13	25.17	7.33	6	0	0.4	675
Egg-substitute	1/4 cup	43	5.5	1.56	0.22	1.5	0	0	0
Eggroll	1 eggroll	173	6.75	4.56	0.89	25	3	0.8	7
Enchilada	1 enchilada	322	10.5	16.89	9.67	30	0	5.8	42
Fish casserole	1 cup	407	18.5	23.78	7.56	26.25	0.75	1.8	70
Fish sandwich	1 sandwich	488	19	26.56	5.89	39.25	3.75	1.5	70
Fish sticks	3 pieces	150	6.75	8.22	2.22	12	0.75	0.7	15
Fish-fried	3 ounces	209	8.75	11.56	2.67	16.25	1.25	0.7	39
Fish-not fried	3 ounces	81	17	1	0.33	7.5	0	0	45
Fish-smoked, pickled	1 ounce	56	6.25	2.33	0.67	0	0	0	14
Grilled cheese sandwich	1 sandwich	442	17	30.67	13.78	23	2.25	1.7	53
Ground beef-lean	3 ounces, 3/4 cup	228	22.5	13.94	5.44	0	0	0	76.5
Ground beef-regular	3 ounces, 3/4 cup	246	20.75	17.56	6.89	0	0	0	80
Ham	3 ounces	124	15.75	6.56	2.33	0	0	0	41

NUTRITIONAL CONTENT OF COMMON FOODS (continued)

NOTE: Carbohydrate value does not include Added Sugar value

	AMOUNT	CAL	PROT (gm)	T FAT (gm)	S FAT (gm)	CARB (gm)	SUG (gm)	FBR (gm)	CHOL (mg)
Ham sandwich	1 sandwich	343	23.5	15.89	7.67	23	2	1.7	60
Hamburger (large) & roll	1 sandwich	594	27.5	33	12.67	33.25	2	0.9	101
Hamburger (lowfat) & roll	1 sandwich	320	22	10	4	35	3.5	1.6	60
Hamburger (small) & roll	1 sandwich	355	22	19.33	8.22	22.25	3	1.7	95
Hot dog	1 hot dog	144	5.75	12.89	5.22	0.25	1.25	0	30
Hot dog and roll	1 sandwich	298	9.25	17.56	6.67	20	2.5	0.7	29
Julienne salad	2 cup	489	47.5	29	13.89	7.5	0	3.5	281
Lamb	3 ounces	169	19.75	11.11	5.33	0	0	0	68
Lentils	1/2 cup	115	7.75	0.22	0	20.25	0	2.8	0
Liver	3 ounces	127	17.75	4.56	2	2.5	0	0	258
Luncheon meat-beef, pork	2 slice	152	8.5	12.22	5.11	0	1	0	36
Luncheon meat-chicken, turkey	2 slice	64	11.5	1.33	0.44	0	0	0	24
Luncheon meat-lean	2 slice	90	8	4	3.11	3	1.5	0	30
Meatloaf	3 ounces	204	16	12	4.44	7	0.25	0.7	104
Miso	1/2 cup	284	14.25	7.33	1.11	39.25	0	7.4	0
Nuts-mixed	3 tablespoons, 26 nuts	170	4.25	13.56	2.22	6.25	0	1.6	0
Pate	1 tablespoon	41	2	3.67	1.44	0	0	0	51
Peanut butter	2 tablespoon	190	9	14.56	2.78	4.5	2	2.4	0
Peanut butter & jelly sandwich	1 sandwich	371	12.25	17.89	3.56	26.5	17.75	3.8	0
Peanuts	3 tablespoons, 32 nuts	164	6.25	12.33	1.78	5.25	0	2.5	0
Peas-black eyed	1/2 cup	100	5.75	0.33	0.11	18.25	0	8.3	0

Peas-black eyed,									
canned	1/2 cup	92	5	0.33	0.11	16.5	0	8.2	0
Pepperoni	3 ounces	440	16.5	41.56	14.56	2.25	2	0	72
Pizza-cheese & vegetable	1 slice	249	12.5	9.22	5.11	30	0.5	4	15
Pizza-cheese topping	1 slice	199	11.5	8.44	4.44	28.25	0.5	2	17
Pizza-cheese,									
meat & vegetable	1 slice	270	16	12.44	6	26.5	0.5	3.5	27
Pizza-French bread	1 slice	410	17.5	19.22	8	39	2	2	35
Pizza-meat topping	1 slice	271	11.5	13	5.33	27	0.5	1.2	20
Pork and beans	1/2 cup	134	6.5	1.89	0.67	17	8	6.6	9
Pork chop	3 ounces	217	23.5	12.89	4.89	0	0	0	70
Pork chop-lean cut	3 ounces	177	0	8	0	0	0	0	0
Pork feet	8 ounces	138	14.5	8.78	3.22	0	0	0	71
Pork rinds	3 ounces	458	51.75	26	10	0	0	0	80
Pork roast	3 ounces	232	23	15	5.56	0	0	0	78
Pork roast-lean cut	3 ounces	180	24	8.33	3.11	0	0	0	78
Pork spareribs	3 ounces	132	10.25	9.89	3.89	0	0	0	41
Pork-fresh, fried	3 ounces	144	11	11.22	4.22	0	0	0	41
Pork-fresh, roasted	4 ounces	164	15.75	10.22	3.89	0	0	0	54
Refried beans	1/2 cup	135	6	1.33	0.56	18	0	6	0
Roast beef sandwich	1 sandwich	353	27.25	14.89	7.33	30.25	2.25	1.7	49
Roast beef-lean cut	3 ounces	156	24.75	5.67	1.89	0	0	0	69
Roast beef-regular	3 ounces	222.5	22.75	13.5	5.11	0	0	0	69.5
Sausage	3 ounces	264	12	22.33	8.67	0	1.5	0	42
Seafood or fish salad	1/2 cup	160	13.5	9.78	2.33	1.75	0.25	0.4	142
Sloppy Joe sandwich	1 sandwich	302	18.5	13.78	6	23.75	2.5	1.4	63
Soybeans-roasted	1/4 cup	205	14.25	10.22	1.56	13.5	0	1.9	0
Steak	3 ounces	191	24	9.33	3.44	0	0	0	71
Steak-lean cut	3 ounces	176	24.5	7.89	3	0	0	0	69
Submarine/hoagy	1 submarine	934	34.5	51.22	12.67	73.25	3.75	3.1	87
Sweet & sour chicken,									

NUTRITIONAL CONTENT OF COMMON FOODS (continued)

NOTE: Carbohydrate value does not include Added Sugar value

	AMOUNT	CAL	PROT (gm)	T FAT (gm)	S FAT (gm)	CARB (gm)	SUG (gm)	FBR (gm)	CHOL (mg)
pork	1 cup	426	17.5	13.89	3.33	23.5	31.75	1.3	83
Taco	1 taco	370	21	18.44	11.11	26.5	0	3.4	57
Taco salad	1-1/2 cup	279	13.5	13.33	6.67	24	0	4.3	44
Tahini	1 tablespoon	92	2.5	7.33	1.11	3.75	0	1.5	0
Tofu	3 ounces	65	6.75	2.33	0.33	2.25	0	1	0
Tostada	1 tostada	325	13.75	13.89	9.67	28	0	7.5	40
Tripe	4 ounces	61	12.5	1.11	0.67	0	0	0	58
Tuna casserole	1 cup, 18 chips	248	16	10.44	3.33	26.75	0	0.7	36
Tuna in oil	1/2 cup	142	22	5.44	1.11	0	0	0	18
Tuna in water	1/2 cup	90	19.25	1.44	0.44	0	0	0	28
Tuna salad	1/2 cup	183	14	9.44	1.89	2.25	0	0.4	13
Tuna sandwich	1 sandwich	364.5	16.25	23	4.11	24.13	2.25	1.7	25
Turkey hot dog	1 hot dog	102	6.5	8	3.11	0.25	1.5	0	51
Turkey sandwich	1 sandwich	355	30.25	14.67	2.67	23.5	2.25	1.7	76
Turkey w/skin	3 ounces	168	24	7.89	2.11	0	0	0	70
Turkey w/out skin	3 ounces	137	25.5	3.22	1	0	0	0	72
Veal	3 ounces	195	25.63	10.17	4	0	0	0	95.5
Veal-lean cut	3 ounces	167	27	5.33	1.67	0	0	0	98
Veal-mixed dish	1 serving	327	28.25	17.78	9.78	9.5	0.75	1.7	137
Venison	3 ounces	147	25.75	3.11	1.22	0	0	0	95
BEVERAGES									
Beer-nonalcoholic	12 fluid ounces, 1-1/2 cup	55	0.75	0	0	11	0	0	0
Coffee	1 cup	5	0.25	0	0	1	0	0	0
Coffee-decaffeinated	1 cup	3	0.25	0	0	0.75	0	0	0

Cola-diet	12 fluid ounces	2	0.25	0	0	0.25	0	0	0
Cola-diet, no caffeine	12 fluid ounces	2	0	0	0	0	0	0	0
Cola-no caffeine	12 fluid ounces	155	0	0	0	0	38.75	0	0
Juice drink	3/4 cup, 1 juice box	106	0	0	0	6.5	19.5	0	0
Mellow Yellow, Mountain Dew	12 fluid ounces	177	0	0	0	0	44	0	0
Noncola-diet, no caffeine	12 fluid ounces	4	0	0	0	0.5	0	0	0
Noncola-no caffeine	12 fluid ounces	157	0	0	0	0	37.75	0	0
Postum	1 teaspoon	12	0	0	0	3	0	1.3	0
Tea-herbal, no caffeine	1 cup	4	0	0	0	0.75	0	0	0
Tea-plain	1 cup	3	0	0	0	0.5	0	0	0
Wine-nonalcoholic	5 fluid ounces	42	0.5	0	0	9.75	0	0	0
BEVERAGES-ALCOHOLIC									
Beer	12 fluid ounces, 1-1/2 cup	145	1	0	0	13.25	0	0	0
Beer-light	12 fluid ounces, 1-1/2 cup	110	1	0	0	7	0	0	0
Chianti	5 fluid ounces	106	0.25	0	0	2.5	0	0	0
Cocktail-mixed drink	1 cocktail	139	0	0	0	1	1.5	0	0
Liqueur	1 glass, 1-1/2 ounce	167	0	0.11	0	8.5	9.5	0	0
Liquor	1 jigger, 1-1/2 fluid ounces	110	0	0	0	0	0	0	0
Vermouth	5 fluid ounces	100	0.25	0	0	1	0	0	0
Wine	5 fluid ounces	104	0.25	0	0	2.5	0	0	0
Wine cooler	12 fluid ounces, 1-1/2 cup	173	0.75	0	0	7.75	9.75	0	0
Wine-light	5 fluid ounces	73	0.5	0	0	1	0	0	0

NUTRITIONAL CONTENT OF COMMON FOODS (continued)

NOTE: Carbohydrate value does not include Added Sugar value

	AMOUNT	CAL	PROT (gm)	T FAT (gm)	S FAT (gm)	CARB (gm)	SUG (gm)	FBR (gm)	CHOL (mg)
FATS/SWEETS/OTHER									
Bacon	1 slice, 1-1/2 tablespoon	37	2.35	3.11	1.16	0.09	0.05	0	8
Bacon bits	1 tablespoon	21	2.5	1.11	0.33	0	0	0	6
Breakfast milk powder	1 packet	130	6	0	0	0	26.25	0.4	0
Brownie	1 square	150	2	6.22	1.67	6.5	15	0.9	10
Butter	1 teaspoon, 1 pat	34.7	0	3.81	2.41	0	0	0	10.7
Cake	1 piece	280	4	11.33	3	15.25	22.25	0.5	56
Candy-chewy	1 ounce	109	0	1	0.56	1.5	20.5	0	0
Candy-chocolate & peanut butter	1 package, 1-1/2 ounces	237	6	13.78	5.89	4	22	2.5	3
Candy-chocolate	1 ounce	150	2	8.22	4.78	2.25	15	0.8	6
Candy-chocolate covered	1 ounce	132	1.25	5.56	2.11	3	13.25	1.5	3
Candy-fudge	1 cube	119	1	4.06	1.06	1.7	17.93	0.4	2.7
Candy-hard	5 pieces	110	0	0	0	0	27.5	0	0
Candy-chocolate & peanut butter	1 package, 1-1/2 ounces	237	6	13.78	5.89	4	22	2.5	3
Candy-chocolate	1 ounce	150	2	8.22	4.78	2.25	15	0.8	6
Candy-chocolate covered	1 ounce	132	1.25	5.56	2.11	3	13.25	1.5	3
Candy-fudge	1 cube	119	1	4.06	1.06	1.7	17.93	0.4	2.7
Candy-hard	5 pieces	110	0	0	0	0	27.5	0	0
Catsup	1 tablespoon	17	0	0	0	2.75	1.5	0.2	0
Cheese puffs	1 cup	160	2	9.78	1.78	16	0	0.3	0
Cheese sauce	1/4 cup	71	3.5	3.56	1.89	6.5	0	0	10

Chili sauce	1 tablespoon	17	0.25	0	0	2.75	1.25	0.9	0
Chip dip	1/4 cup	120	2	10	6	4	0.5	0	40
Coffee whitener	1 tablespoon	22	0	2.11	1.33	0	1	0	0
Cookies-fig bars	2 bar	100	1	1.78	0.44	10.5	10.5	1.2	1
Cookies-nonfat	2 cookies	75	1	0	0	17.5	6.5	0.6	0
Cookies-oatmeal raisin	3 cookies	195	2.25	8.11	1.89	13.5	13.5	1.4	1
Cookies-other	3 cookies	180	1.5	8.11	2.89	10	15	0.3	3
Corn chips	1 cup	152	2	9.44	1.33	16	0	1.3	0
Cream cheese	2 tablespoon	106	2.5	10.22	6.67	1	0	0	34
Cream cheese-light	2 tablespoon	80	3	7	4	1	0	0	25
Cream-coffee, half & half	1 tablespoon	25	0.5	2.22	1.44	0.5	0	0	8
Cream-whipped	1 tablespoon	15	0.25	1.33	1.11	0.25	0.25	0	2
Cupcakes	1 cupcake	140	1.25	4.56	1.89	8.5	15	0.7	12
Danish	1 danish	252	4.5	11.67	3.67	10.25	19	0.7	14
Danish-nonfat	1 danish	90	2	0	0	20	9.75	0.2	0
Dessert topping-no sugar	1 tablespoon	5	0	0.56	0.44	0	0	1	4
Diet bar	1 bar	120	2	4	1.44	19	9.5	3	1
Doughnut or sweet roll	1 doughnut, 1 sweetroll	220	3.75	9.89	3.33	29.5	18	1	5
Frozen desserts-nonfat	1/2 cup	100	2	0.22	0.11	23.5	9.5	0.4	1
Frozen yogurt cone-lowfat	1 serving	105	4	1	0.56	22	13	0.1	3
Frozen yogurt sundae-lowfat	1 sundae	240	6	3	2.33	50.5	43	0.8	6
Gelatin	1/2 cup	105	2	1	1	23	22	0	0
Gelatin-sugar free	1/2 cup	8	1.5	0	0	0	0	0	0
Granola bars	1 bar	133	2	6	2	18.25	13	0.6	0
Gravy	1/4 cup	30	1	1.44	0.56	2.25	0.5	0.1	1
Hollandaise sauce	1/4 cup	230	2.25	23.22	8.44	2.5	0	0	140
Honey	2 teaspoon	42	0	0	0	0	10.5	0	0
Hot cocoa mix	1 envelope	110	1.5	2.78	1.56	3.5	16	1.1	2
Ice cream bar	1 bar	172	2	11.78	7.11	6	11.5	0.3	17

NUTRITIONAL CONTENT OF COMMON FOODS (continued)

NOTE: Carbohydrate value does not include Added Sugar value

	AMOUNT	CAL	PROT (gm)	T FAT (gm)	S FAT (gm)	CARB (gm)	SUG (gm)	FBR (gm)	CHOL (mg)
Jam or jelly	2 teaspoon	35	0	0	0	1	7.5	0.1	0
Lard	1 tablespoon	115	0	12.22	5	0	0	0	12
Margarine-stick	1 teaspoon, 1 pat	33.7	0	3.52	0.67	0	0	0	0
Margarine-stick, light	1 teaspoon, 1 pat	20	0	2.22	0.33	0	0	0	0
Margarine-tub	1 tablespoon	101	0	8.89	2	0	0	0	0
Margarine-tub, light	1 tablespoon	50	0	5.89	1	0	0	0	0
Marshmallows	2 piece	47	0.25	0	0	0	11.75	0	0
Mayonnaise	1 tablespoon	100	0.25	11	1.89	0.25	0.25	0	8
Mayonnaise-light	1 tablespoon	48	0.25	4.56	1	0.75	1	0	5
Mayonnaise-nonfat	1 tablespoon	12	0	0	0	3	3	0	0
Meal replacement bar	1 serving	270	11	14	5	24	22.5	0	0
Milkshake	10 fluid ounces, 1-1/4 cup	368	10	12.78	8.22	26.5	19.25	0.5	54
Milkshake-lowfat	1 serving	320	10.75	1.33	0.56	66	44.75	0	10
Miracle Whip	1 tablespoon	64	0	5.89	0.89	2.5	0.5	0	5
Miracle Whip-nonfat	1 tablespoon	20	0	0	0	5	5	0	0
Molasses	1 tablespoon	55	0	0	0	0	14.5	0	0
Mustard	1 teaspoon	6	0.25	0.33	0	0.25	0	0	0
Noncola-diet	12 fluid ounces, 1-1/2 cup	2	0	0	0	0.5	0	0	0
Nutrasweet-Equal	1 packet	4	0.5	0	0	0	0.25	0	0
Oil	1 tablespoon	120	0	12.78	1.89	0	0	0	0
Pickles-dill	2 spear	7	0	0.11	0	1	0.75	0.9	0
Pickles-sweet	1 pickle, 3 slices	18	0	0	0	0.5	4	0.3	0
Pie-custard or cream	1 slice	346	6.75	13.11	6.22	20.75	25.75	1	125
Pie-fruit	1 slice	405	4	16.22	5.33	34.75	25.25	4	6

Pie-pecan	1/6 of 9" pie	575	7	29.67	5.67	33	37.25	2.2	100
Popcorn	1 cup	32	0.75	1.44	0.44	5.5	0	0.8	0
Popsicle	1 popsicle	50	0	0	0	0	13	0	0
Potato chips	1 cup, 20 chips	150	2	9.67	2	15	0	1.4	0
Pretzels	2/3 cup	110	2.75	0.89	0.22	21.75	1	0.9	0
Pudding	1/2 cup	150	4.5	2.22	1.44	10	18	0	9
Pudding-diet	1/2 cup	90	4	2.44	1.56	13	0	0.4	9
Relish	2 tablespoon	35	0	0	0	1.25	7.5	0.2	0
Saccharin	1 packet	2	0	0	0	0	0	0	0
Salad dressing	1 tablespoon	80	0	8.22	1.33	0.25	0.25	1	0
Salad dressing-light	1 tablespoon	16	0.25	0.33	0.11	0.75	0.5	0.3	0
Salad dressing-no oil	1 tablespoon	12	0	0	0	2.5	0	0.7	0
Salad dressing-nonfat	1 tablespoon	16	0	0	0	3	3	0	0
Salt	4 shakes	0	0	0	0	0	0	0	0
Sherbet	1/2 cup	136	1	1.89	1.22	6	23.5	0	7
Soft drinks	12 fluid ounces, 1-1/2 cup	156	0	0	0	0	39.5	0	0
Soup-beef or chicken	1 cup	74	4.25	2.22	0.67	8.5	0	1	7
Soup-bouillon, broth	1 cube, packet(s), 1 cup	9	0.75	0.22	0.11	0.25	1	0	1
Soup-broth based, no salt	1 cup	135	5.75	3.89	0.78	16.5	0	2.8	0
Soup-cream, chowder	1 cup	140	5.5	6.11	2.89	14	0	0.9	22
Soup-low salt	1 cup	110	4	3	1	12	0	0.5	2
Soup-miso	1 cup	152	4.5	6.44	0.89	19	0	3	0
Sour cream	2 tablespoon	52	0.5	5.11	3.11	1	0	0	10
Sour cream-imitation	2 tablespoon	50	1.5	4.67	4	1.5	0	0	2
Sour cream-nonfat	2 tablespoon	16	2	0	0	2	0	0	0
Soy sauce	1 tablespoon	10	1.25	0	0	1.25	0	0	0
Steak/Worcestershire sauce	1 tablespoon	11	0	0	0	1	1.75	0	0
Sugar	1 teaspoon	15	0	0	0	0	3.75	0	0

NUTRITIONAL CONTENT OF COMMON FOODS (continued)

NOTE: Carbohydrate value does not include Added Sugar value

	AMOUNT	CAL	PROT (gm)	T FAT (gm)	S FAT (gm)	CARB (gm)	SUG (gm)	FBR (gm)	CHOL (mg)
Sunflower seeds	2 tablespoon	116	3.75	10.78	1.11	4	0	1.7	0
Syrup-pancake, table	2 tablespoon	110	0	0	0	0	27.5	0	0
Tortilla chips	1 cup, 10 chips	95	1.25	4.67	1.33	12	0	0.9	0
White sauce	1/4 cup	99	2.5	5.67	2.22	6	0	0.1	8

PART TWO

Nutritional Content of Fast Foods

NUTRITIONAL CONTENT OF FAST FOODS

Burger King

	Serving Size (g)	Calories	Calories from Fat	Total Fat (g)	Saturated Fat (g)	Cholesterol (mg)	Sodium (mg)	Total Carbohydrate (g)	Dietary Fiber (g)	Sugars (g)	Protein (g)	% Daily Value‡ Vitamin A	Vitamin C	Calcium	Iron
Whopper® Sandwich	270	640	350	39	11	90	870	45	3	8	27	10	15	8	25
Whopper® with Cheese Sandwich	294	730	410	46	16	115	1300	46	3	8	33	15	15	25	25
Double Whopper® Sandwich	351	870	500	56	19	170	940	45	3	8	46	10	15	8	40
Double Whopper® with Cheese Sandwich	375	960	570	63	24	195	1360	46	3	8	52	15	15	25	40
Whopper JR.® Sandwich	168	420	220	24	8	60	570	29	2	5	21	4	8	6	20
Whopper JR.® with Cheese Sandwich	180	460	250	28	10	75	780	29	2	5	23	8	8	15	20
Hamburger	129	330	140	15	6	55	570	28	1	4	20	2	0	4	15
Cheeseburger	142	380	170	19	9	65	780	28	1	5	23	6	0	15	15
Double Cheeseburger	213	600	320	36	17	135	1040	29	1	5	41	8	0	20	25
Double Cheeseburger with Bacon	221	640	350	39	18	145	1220	29	1	5	44	8	0	20	25
Sandwiches/Side Orders															
BK Big Fish Sandwich	255	720	390	43	8	60	1090	59	2	4	25	2	2	6	20
BK Broiler® Chicken Sandwich	248	540	260	29	6	80	480	41	2	3	30	4	10	4	30

Chicken Sandwich	229	700	380	43	9	60	1400	54	2	4	26	*	*	10	20
Chicken Tenders® (6 piece)	88	250	110	12	3	35	530	14	2	0	16	*	*	*	4
Broiled Chicken Salad†	302	200	90	10	5	60	110	7	3	4	21	100	25	15	20
Garden Salad†	215	90	45	5	3	15	110	7	3	0	6	110	50	15	6
Side Salad†	133	50	25	3	2	5	55	4	2	0	3	50	20	6	2
French Fries (medium, salted)	116	400	180	20	5	0	240	43	3	0	5	*	4	*	6
Onion Rings	124	310	130	14	2	0	810	41	5	6	4	*	*	*	*
Dutch Apple Pie	113	310	140	15	3	0	230	39	2	22	3	*	10	*	8
Drinks															
Vanilla Shake (medium)	284	310	60	7	4	20	230	53	1	47	9	6	6	30	*
Chocolate Shake (medium)	284	310	60	7	4	20	230	54	3	48	9	6	*	20	10
Chocolate Shake (medium, syrup added)	341	460	70	7	4	20	300	87	1	22	11	6	6	30	*
Strawberry Shake (medium, syrup added)	341	430	60	7	4	20	260	83	1	47	9	6	6	30	*
Coca-Cola® Classic (medium)	22 (fl oz)	260	0	0	0	0	@	70	0	70	0	*	*	*	*
Diet Coke® (medium)	22 (fl oz)	1	0	0	0	0	@	<1	0	<1	0	*	*	*	*
Sprite® (medium)	22 (fl oz)	260	0	0	0	0	@	66	0	66	0	*	*	*	*
Tropicana® Orange Juice	311	140	0	0	0	0	0	33	0	28	2	0	100	0	0
Coffee	355	5	0	0	0	0	5	1	0	0	0	*	*	*	*
Milk, 2% Low fat	244	120	40	5	3	20	120	12	0	0	8	10	4	30	*
Breakfast															
Croissan'wich® with Bacon, Egg and Cheese	118	350	220	24	8	225	790	18	<1	2	15	8	*	15	10
Croissan'wich® with Sausage, Egg and Cheese	159	530	370	41	14	255	1000	21	<1	2	20	8	*	15	15
Croissan'wich® with Ham, Egg and Cheese	144	350	200	22	7	230	1390	19	<1	2	18	8	*	15	10

NUTRITIONAL CONTENT OF FAST FOODS (continued)

Burger King

	Serving Size (g)	Calories	Calories from Fat	Total Fat (g)	Saturated Fat (g)	Cholesterol (mg)	Sodium (mg)	Total Carbohydrate (g)	Dietary Fiber (g)	Sugars (g)	Protein (g)	% Daily Value‡ Vitamin A	Vitamin C	Calcium	Iron
French Toast Sticks	141	500	240	27	7	0	490	60	1	11	4	•	•	6	15
Hash Browns	71	220	110	12	3	0	320	25	2	0	2	10	8	*	2
A.M. Express® Grape Jam	12	30	0	0	0	0	0	7	0	6	0	0	0	0	0
A.M. Express® Strawberry Jam	12	30	0	0	0	0	5	8	0	5	0	0	0	0	0

‡Percent Daily Values are based on a 2,000 calorie diet.
*Contains less than 2% of the daily value of this nutrient.

†Without dressing
@Depends on the water supply.
#Regular Italian Dressing: 150 calories, 16 grams (g) fat.

Source: © Burger King Corporation. The nutritional information on this brochure was updated in December, 1994. For the latest nutritional information, please check your local Burger King® restaurant. "Nutritional Information Chart" reprinted by permission.

Domino's Pizza

	Serving Size (g)	Calories (kcal)	Calories from Fat (kcal)	Protein (g)	Carbohydrates (g)	Sugars (g)	Dietary Fiber (g)	Fat-total (g)	Saturated Fat (g)	Cholesterol (mg)	Ash (g)	Water (g)	Vitamin A (IU)	Vitamin C (mg)	Calcium (mg)	Iron (mg)	Sodium (mg)
12" Hand-Tossed																	
Cheese	Per 2 of 8 slices (147.4 g)	344.2	90	14.8	50.0	1.0	2.4	9.5	4.4	19.1	3.4	68.7	456.2	2.6	277.8	4.0	980.5
Pepperoni	Per 2 of 8 slices (159.4 g)	406.2	140	17.5	50.2	1.0	2.5	15.1	6.6	32.1	4.1	71.5	467.6	2.7	282.3	4.3	1179.0
X-Tra Cheese & Pepperoni	Per 2 of 8 slices (175.3 g)	455.0	170	20.9	50.5	1.0	2.5	18.8	8.6	41.7	4.6	79.2	578.7	2.7	411.8	4.5	1304.0
Ham	Per 2 of 8 slices (160.7 g)	361.6	90	17.2	50.3	1.2	2.4	10.2	4.6	26.1	4.0	78.1	456.5	2.7	279.3	4.2	1143.0
Italian Sausage & Mushroom	Per 2 of 8 slices (176 g)	402.4	120	17.5	52.2	1.3	2.9	13.9	6.1	30.5	4.2	87.2	483.0	3.2	286.5	4.5	1151.0
Veggie*	Per 2 of 8 slices (176 g)	360.0	90	15.2	51.7	1.2	3.0	10.4	4.5	19.1	3.6	93.9	493.5	12.7	285.9	4.4	1028.8
12" Thin Crust																	
Cheese	1/3 pizza (140.9 g)	364.3	140	16.1	40.1	2.2	1.9	15.5	6.3	25.5	3.7	64.2	567.4	3.5	422.3	1.5	1012.0
Pepperoni	1/3 pizza (156.7 g)	447.1	210	19.7	40.4	2.3	2.0	23.0	9.2	42.8	4.7	67.8	582.6	3.5	428.4	1.8	1277.0

NUTRITIONAL CONTENT OF FAST FOODS (continued)

Domino's Pizza

	Serving Size (g)	Calories (kcal)	Calories from Fat (kcal)	Protein (g)	Carbohydrates (g)	Sugars (g)	Dietary Fiber (g)	Fat-total (g)	Saturated Fat (g)	Cholesterol (mg)	Ash (g)	Water (g)	Vitamin A (IU)	Vitamin C (mg)	Calcium (mg)	Iron (mg)	Sodium (mg)
X-Tra Cheese & Pepperoni	1/3 pizza (177.7 g)	512.3	250	24.2	40.8	2.3	2.0	28.0	12.0	55.6	5.3	77.8	731.0	3.5	601.1	2.1	1443.0
Ham	1/3 pizza (158.5 g)	387.7	150	19.3	40.5	2.5	1.9	16.5	6.6	34.8	4.4	76.5	567.9	3.6	424.3	1.7	1229.0
Italian Sausage & Mushroom	1/3 pizza (178.7 g)	442.3	190	19.8	43.1	2.6	2.5	21.4	8.6	40.7	4.8	88.5	603.5	4.2	434.0	2.1	1240.0
Veggie*	1/3 pizza (178.7 g)	385.7	150	16.7	42.5	2.5	2.6	16.7	6.5	25.5	3.9	97.5	617.5	16.9	433.3	2.0	1076.0
12" Deep Dish Cheese	Per 2 of 8 slices (205.3 g)	559.7	210	23.5	63.2	4.3	3.2	23.8	9.0	31.5	4.4	92.2	763.9	3.0	451.4	4.8	1184.0
Pepperoni	Per 2 of 8 slices (218 g)	621.8	260	26.2	63.4	4.4	3.2	29.4	11.2	44.5	5.1	95.7	775.3	3.0	455.9	5.0	1383.0
X-Tra Cheese & Pepperoni	Per 2 of 8 slices (234.9 g)	670.7	300	29.6	63.7	4.4	3.2	33.1	13.3	54.0	5.5	104.3	886.6	3.0	585.5	5.2	1508.0
Ham	Per 2 of 8	577.2	220	25.9	63.5	4.5	3.2	24.5	9.3	38.4	4.9	102.3	764.3	3.1	452.9	4.9	1347.0

	slices (219.4 g)															
Italian Sausage & Mushroom	Per 2 of 8 slices (235.7 g)	618.2	250	26.2	65.5	4.6	3.7	28.2	10.8	42.9	5.1	112.4	791.0	3.5	460.2	5.2	1356.0
Veggie*	Per 2 of 8 slices (235.7 g)	575.8	220	24.0	65.0	4.5	3.7	24.7	9.2	31.5	4.5	119.1	801.5	13.0	459.6	5.1	1233.0

*Veggie includes fresh mushrooms, onions, green peppers, & ripe olives.

Data is based on minimal portioning requirements. Nutrient values may vary slightly by location and supplier base.

Source: ©1994 Domino's Pizza, Inc. Reprinted by permission of Domino's Pizza, Inc.

NUTRITIONAL CONTENT OF FAST FOODS (continued)

Kentucky Fried Chicken

	Serving Size (g)	Calories	Calories from Fat	Total Fat (g)	% Daily Value	Saturated Fat (g)	% Daily Value	Cholesterol (mg)	% Daily Value	Sodium (mg)	% Daily Value	Carbohydrates (g)	% Daily Value	Dietary Fiber (g)	% Daily Value	Sugars (g)	Protein (g)	Vitamin A	Vitamin C	Calcium	Iron
																		% Daily Value			
Colonel's Rotisserie Gold® Chicken Quarter—Breast and Wing	6.2 oz	335	168	18.7	29	54	27	157	52	1104	46	1	—	—	—	—	40	<2	<2	<2	<2
Colonel's Rotisserie Gold® Chicken Quarter—Breast and Wing with skin and wing removed by customer	4.1 oz	199	53	5.9	9	1.7	9	97	32	667	28	0	—	—	—	—	37	<2	<2	<2	<2
Original Recipe® Chicken Breast	4.8 oz	360	180	20	31	5	27	115	38	870	36	12	4	1	4	0	33	—	—	4	6
Extra Tasty Crispy™ Chicken Breast	5.9 oz	470	250	28	42	7	35	80	27	930	39	25	8	1	4	0	31	—	—	4	6
Hot & Spicy Chicken Breast	6.5 oz	530	310	35	54	8	42	110	36	1110	46	23	8	2	9	0	32	—	—	4	6
Mashed Potatoes with Gravy	4.2 oz	109	45	5	8	<1	<5	<1	<1	386	16	16	5	2	8	<1	1	—	—	—	—
Cornbread	2.0 oz	228	117	13	20	2	11	42	14	194	8	25	8	1	4	10	3	1	0	6	4
Corn on the Cob	5.3 oz	222	104	12	18	2	9	0	0	76	3	27	9	8	31	4	4	4	3	0	2

Source: Courtesy of Kentucky Fried Chicken (KFC) Corp.Sandwiches

	Serving Size (g)	Calories	Calories from Fat	Total Fat (g)	% Daily Value	Saturated Fat (g)	% Daily Value	Cholesterol (mg)	% Daily Value	Sodium (mg)	% Daily Value	Carbohydrates (g)	% Daily Value	Dietary Fiber (g)	% Daily Value	Sugars (g)	Protein (g)	Vitamin A	Vitamin C	Calcium	Iron
																		% Daily Value			
Sandwiches																					
Hamburger	106 g	250	80	9	14	3.5	15	40	12	490	20	32	11	2	7	1	12	4	4	10	15
Cheeseburger	119 g	300	120	13	20	5	25	50	17	730	30	33	11	2	7	1	15	8	4	20	15
Quarter Pounder	166 g	400	180	20	31	8	40	85	28	640	27	35	12	2	8	2	23	4	6	15	20
Quarter Pounder with Cheese	191 g	490	250	27	43	10	55	115	38	1090	46	36	12	2	8	3	28	15	6	30	20
McLean Delux™	206 g	320	90	10	15	4	20	60	20	670	28	35	12	3	10	3	22	10	10	15	20
McLean Delux™ with Cheese	219 g	370	130	14	22	5	25	75	25	890	37	35	12	3	10	3	24	15	10	20	20
Big Mac®	215 g	490	240	27	40	9	45	90	33	890	37	49	13	2	9	1	24	6	2	25	20
Filet-O-Fish®	141 g	370	160	18	28	4	20	50	17	730	30	38	13	2	7	0	14	2	*	15	20
McGrilled Chicken Sandwich	240 g	400	110	12	18	4	20	80	27	680	28	42	14	1	4	2	31	8	15	15	8
McChicken® Sandwich	182 g	490	270	30	45	6	28	45	15	760	32	43	14	3	11	1	16	4	2	15	15
Chicken Fajita	82 g	190	70	8	12	2	10	35	12	310	13	20	7	1	5	1	11	2	10	8	4
French Fries																					
Small French Fries	68 g	220	110	12	18	2.5	13	0	0	110	5	26	9	2	9	0	3	*	15	*	2
Large French Fries	122 g	400	200	22	34	5	25	0	0	200	8	46	16	4	16	1	6	*	25	*	6
Salads																					
Chef Salad	265 g	170	80	9	14	4	20	110	37	400	17	8	3	2	9	4	17	100	35	15	8
Chunky Chicken Salad	255 g	150	35	4	6	1	5	80	27	230	10	7	2	2	9	3	25	170	45	4	6
Garden Salad	189 g	50	20	2	3	0.5	3	65	22	70	3	6	2	2	10	4	4	90	35	4	8
Breakfast																					
Egg McMuffin®	135 g	280	100	11	17	4	20	235	78	710	30	28	9	1	6	2	18	10	*	25	15
Sausage McMuffin®	135 g	350	180	20	31	7	35	60	20	770	32	27	9	1	6	2	15	4	*	20	15

NUTRITIONAL CONTENT OF FAST FOODS (continued)

McDonalds

% Daily Value (Vitamin A, Vitamin C, Calcium, Iron)

	Serving Size (g)	Calories	Calories from Fat	Total Fat (g)	% Daily Value	Saturated Fat (g)	% Daily Value	Cholesterol (mg)	% Daily Value	Sodium (mg)	% Daily Value	Carbohydrates (g)	% Daily Value	Dietary Fiber (g)	% Daily Value	Sugars (g)	Protein (g)	Vitamin A	Vitamin C	Calcium	Iron
Sausage McMuffin® with Egg	159 g	430	230	25	38	8	40	270	90	920	38	27	9	1	6	2	21	10	*	25	20
English Muffin	58 g	170	35	4	6	1	5	0	0	290	12	26	9	2	6	2	5	2	*	15	8
Sausage Biscuit	118 g	420	250	28	43	8	40	45	15	1040	43	32	11	0	0	0	12	*	*	8	10
Sausage Biscuit with Egg	175 g	520	310	34	51	11	50	270	87	1240	50	33	11	0	0	3	19	6	*	10	20
Bacon, Egg & Cheese Biscuit	153 g	440	240	26	40	8	40	240	80	1220	51	33	11	0	0	1	15	10	*	20	15
Biscuit	75 g	260	120	13	20	3	15	1	0	730	30	32	11	1	5	3	5	*	*	8	8
Sausage	43 g	160	140	15	23	5	25	45	15	310	13	0	0	0	0	0	7	*	*	*	4
Scrambled Eggs (2)	100 g	140	90	10	15	3	15	425	142	290	12	1	0	0	0	0	12	10	*	6	10
Hash Browns	53 g	130	70	7	11	1	5	0	0	330	14	15	5	1	4	0	1	*	2	*	*
Hotcakes (plain)	126 g	250	35	4	8	1	5	10	3	570	2	43	14	0	0	10	8	*	*	10	11
Hotcakes with Syrup & Margarine (2 pats)	174 g	410	80	9	18	1.5	10	10	3	670	28	75	25	0	0	30	8	4	*	10	10
Desserts/Shakes																					
Vanilla Lowfat Frozen Yogurt Cone	84 g	110	10	1	2	0.5	3	5	2	80	3	22	8	0	0	19	4	2	*	10	*
Strawberry Lowfat Frozen Yogurt Sundae	172 g	210	10	1	2	0.5	3	5	2	100	4	49	16	0	0	48	6	4	2	20	*
Hot Fudge Lowfat Frozen Yogurt Sundae	168 g	240	25	3	5	2	10	5	2	170	7	50	16	0	0	44	7	4	*	25	2
Baked Apple Pie	84 g	280	140	15	23	2	10	0	0	90	4	35	11	1	6	15	3	*	*	2	6
McDonaldland Cookies®	1pkg.	290	80	9	14	1	5	0	0	300	13	47	16	1	5	21	4	*	*	*	10

Chocolaty Chip Cookies	1 pkg.	330	140	15	23	4	20	5	2	280	12	42	14	2	7	15	4	*	*	2	10
Vanilla Shake, small (16 fl oz)	480 ml	310	45	5	8	3	15	25	8	170	7	55	17	0	0	51	12	4	4	35	2
Chocolate Shake, small (16 fl oz)	480 ml	350	60	6	9	4	20	25	8	240	10	62	19	0	0	58	13	4	4	35	6
Strawberry Shake, small (16 fl oz)	480 ml	340	45	5	8	3	15	25	8	170	7	63	20	0	0	59	12	4	4	35	2

Source: ©1994 McDonald's Corporation. Rev. 3/94, McD 2–1023, –030.

NUTRITIONAL CONTENT OF FAST FOODS (continued)

Taco Bell

	Serving Size (g)	Calories	% Calories Reduction	Calories from Fat (g)	Total Fat	% Fat Reduction	Cholesterol (mg)	Protein (g)	Carbohydrates (g)
Taco	78	180	—	100	11	—	30	10	11
LIGHT Taco	78	140	22	50	5	55	20	11	11
Soft Taco	99	220	—	100	11	—	30	12	19
LIGHT Soft Taco	99	180	18	50	5	55	25	13	19
Taco Supreme™	106	230	—	140	15	—	45	11	12
LIGHT Taco Supreme™	106	160	30	50	5	67	20	13	14
Soft Taco Supreme®	128	270	—	140	15	—	45	13	21
LIGHT Soft Taco Supreme®	128	200	26	50	5	67	25	14	23
Bean Burrito	198	390	—	110	12	—	5	13	58
LIGHT Bean Burrito	198	330	15	60	6	50	5	14	55
7-Layer Burrito (no longer offered)	276	540	—	210	23	—	20	17	65
LIGHT 7-Layer Burrito	276	440	19	80	9	61	5	19	67
Burrito Supreme	248	440	—	170	19	—	45	18	50
LIGHT Burrito Supreme	248	350	20	70	8	58	25	20	50
Taco Salad	535	860	—	500	55	—	80	32	64
LIGHT Taco Salad (with chips)	535	680	21	230	25	55	50	35	81
(without chips)	464	330	—	80	9	—	50	30	35

LIGHT = 50% less fat per serving.

	Serving Size (g)	Weight (g)	Calories	Calories from Fat	Total Fat (g)	Saturated (g)	Cholesterol (mg)	Sodium (mg)	Total Carbohydrates (g)	Dietary Fiber (g)	Sugars (g)	Protein (g)	Vitamin A (% Daily Value)	Vitamin C (% Daily Value)	Calcium (% Daily Value)	Iron (% Daily Value)
Sandwiches																
Plain Single	1 ea.	133	300	140	16	6	65	460	31	2	5	25	0	0	10	20
Single with everything	1 ea.	219	420	180	20	7	70	810	32	3	9	26	6	10	10	30
Big Bacon Classic	1 ea.	287	610	290	33	13	105	1510	45	3	11	36	15	25	25	35
Jr. Hamburger	1 ea.	117	270	90	10	3	30	560	34	2	7	15	2	2	10	20
Jr. Cheeseburger	1 ea.	129	320	120	13	6	45	770	34	2	7	17	6	2	15	20
Jr. Bacon Cheeseburger	1 ea.	170	410	190	21	8	60	910	34	2	7	22	8	15	15	25
Jr. Cheeseburger Deluxe	1 ea.	179	360	150	16	6	45	840	36	3	8	18	10	10	15	20
Hamburger, Kids' Meal	1 ea.	111	270	90	10	3	30	560	33	2	7	15	2	0	10	20
Cheeseburger, Kids' Meal	1 ea.	123	320	120	13	6	45	770	33	2	7	17	6	0	15	20
Grilled Chicken Sandwich	1 ea.	177	290	60	7	1.5	55	720	35	2	8	24	4	10	10	15
Breaded Chicken Sandwich	1 ea.	208	440	160	18	3	60	840	44	2	6	26	4	10	10	80
Chicken Club Sandwich	1 ea.	220	500	200	23	5	70	1090	44	2	7	30	4	15	10	80
French Fries																
Small	3.2 oz	91	260	120	13	2.5	0	85	33	3	0	3	0	8	2	4
Medium	4.6 oz	130	380	170	19	4	0	120	47	5	0	5	0	10	2	6
Biggie	5.6 oz	159	460	200	23	5	0	150	58	6	0	6	0	15	2	8
Baked Potato																
Plain	10 oz	284	310	0	0	0	0	25	71	7	5	7	0	60	2	20

NUTRITIONAL CONTENT OF FAST FOODS (continued)

Wendy's

	Serving Size (g)	Weight (g)	Calories	Calories from Fat	Total Fat (g)	Saturated (g)	Cholesterol (mg)	Sodium (mg)	Total Carbohydrates (g)	Dietary Fiber (g)	Sugars (g)	Protein (g)	Vitamin A (% Daily Value)	Vitamin C (% Daily Value)	Calcium (% Daily Value)	Iron (% Daily Value)
Bacon & Cheese	1 ea.	380	540	160	18	4	20	1430	78	7	5	17	10	60	20	25
Broccoli & Cheese	1 ea.	411	470	120	14	3	5	470	80	9	6	9	35	120	20	25
Cheese	1 ea.	383	570	210	23	9	30	640	78	7	5	14	15	60	40	20
Chili & Cheese	1 ea.	439	620	220	24	9	40	780	83	9	7	20	20	60	35	35
Sour Cream & Chives	1 ea.	314	380	60	6	4	15	40	74	8	6	8	30	80	8	25
Sour Cream	1 pkt.	28	60	50	6	4	10	15	1	0	1	1	4	0	4	0
Whipped Margarine	1 pkt.	14	60	60	7	1	0	110	0	0	0	0	10	0	0	0
Chili																
Small	8 oz	227	210	60	7	2.5	40	800	21	5	5	15	8	6	8	15
Large	12 oz	340	310	90	10	4	45	1190	32	7	8	23	10	10	10	25
Cheddar Cheese, shredded	2 T.	17	70	50	6	3	15	110	1	0	0	4	4	0	10	0
Saltine Crackers	2 ea.	6	25	5	0.5	0	0	80	4	0	0	0	0	0	0	2
Chicken Nuggets																
6 piece	6	94	280	180	20	5	50	600	12	0	N/A	14	0	0	2	4
Barbeque Sauce	1 pkt.	28	50	0	0		0	100	11	N/A	N/A	1	6	0	0	4
Honey	1 pkt.	14	45	0	0	0	0	0	12	0	12	0	0	0	0	0
Sweet & Sour Sauce	1 pkt.	28	45	0	0	0	0	55	11	N/A	N/A	0	0	0	0	2
Sweet Mustard Sauce	1 pkt.	28	50	10	1	0	0	140	9	N/A	N/A	1	0	0	0	0

PART THREE

Energy Costs of Various Activities

ENERGY COSTS OF VARIOUS ACTIVITIES

Activity	Body Weight 120 Pounds (54 Kilograms) Kcal/Hour	150 Pounds (68 Kilograms) Kcal/Hour	180 Pounds (82 Kilograms) Kcal/Hour
Aerobics, high impact class	381	476	571
Aerobics, low impact class	272	340	408
Aerobics, step, 8 inch bench, 120 spm	400	500	600
Backpacking	381	476	571
Baseball, catch	136	170	204
Baseball, game	272	340	408
Basketball, game	435	544	653
Basketball, shooting baskets	245	306	367
Bicycling, 10-11.9 mph, leisure	327	408	490
Bicycling, 16-19 mph, racing	653	816	980
Bicycling, BMX or mountain biking	463	578	694
Bicycling, stationary, moderate	381	476	571
Bicycling, stationary, vigorous	571	714	857
Billiards	136	170	204
Bowling	163	204	245
Boxing, punching bag	327	408	490
Calisthenics, light or moderate	245	306	367
Calisthenics, vigorous	435	544	653
Canoeing, on camping trip	218	272	327
Catch, throwing/catching ball	206	257	309
Fishing, general	218	272	327
Football, catch	136	170	204
Football, competitive	490	612	735
Golf, general	245	306	367
Gymnastics	218	272	327
Handball, general	653	816	980
Hang gliding	190	238	286
Health club exercise, general	299	374	449
Hiking, cross-country	327	408	490
Hockey, ice	435	544	653
Hockey, street	438	547	656
Horseback riding, general	218	272	327
Hunting, general	272	340	408
Ice skating, general	381	476	571
Jet-skiing	327	408	490
Judo	544	680	816
Jumping rope, 125 per minute	578	722	867
Karate	544	680	816
Kayaking	272	340	408
Kick boxing	544	680	816
Kickball	381	476	571

Activity	Body Weight 120 Pounds (54 Kilograms) Kcal/Hour	150 Pounds (68 Kilograms) Kcal/Hour	180 Pounds (82 Kilograms) Kcal/Hour
Lacrosse	435	544	653
Motorcycle riding	170	212	255
Ping pong	218	272	327
Racquetball, competitive	544	680	816
Rollerskating	381	476	571
Rowing, canoeing, pleasure	190	238	286
Rowing, crew or sculling, competition	653	816	980
Rowing, sliding seat, fast	509	637	764
Rowing, sliding seat, slow	278	347	416
Rugby	544	680	816
Running, 12 mph, 5 minute mile	944	1180	1416
Running, 5 mph, 12 minute mile	435	544	653
Running, 6.7 mph, 9 minute mile	599	748	898
Running, 8.6 mph, 7 minute mile	762	952	1143
Running, cross country	490	612	735
Sailing, competition	272	340	408
Sailing, pleasure	163	204	245
Scuba diving	381	476	571
Shooting, pistol or trap, standing	136	170	204
Skiing, cross country, 2.5 mph	381	476	571
Skiing, cross country, 8 mph and up	762	952	1143
Skiing, downhill, moderate	327	408	490
Sky diving	190	238	286
Sledding	381	476	571
Snowmobiling	190	238	286
Soccer, casual	381	476	571
Soccer, competitive	544	680	816
Softball	272	340	408
Squash	653	816	980
Stair climbing, no load	287	359	431
Stair master, general	327	408	490
Surfing	163	204	245
Swimming, freestyle, fast laps	544	680	816
Swimming, freestyle, slow laps	435	544	653
Swimming, leisurely, general	327	408	490
Tai chi	218	272	327
Tennis, doubles	327	408	490
Tennis, singles	435	544	653
Treadmill	327	408	490
Volleyball, beach	435	544	653
Volleyball, competitive	218	272	327
Volleyball, noncompetitive	163	204	245
Walking, 2 mph, firm surface	136	170	204

ENERGY COSTS OF VARIOUS ACTIVITIES (continued)

Activity	Body Weight 120 Pounds (54 Kilograms) Kcal/Hour	150 Pounds (68 Kilograms) Kcal/Hour	180 Pounds (82 Kilograms) Kcal/Hour
Walking, 3 mph, firm surface	190	238	286
Walking, 4 mph, firm surface	218	272	327
Walking, backpacking	381	476	571
Walking, racing	354	442	531
Water aerobics, water calisthenics	218	272	327
Water skiing	327	408	490
Water volleyball	163	204	245
Weight lifting, light workout	163	204	245
Weight lifting, power lifting, vigorous	327	408	490
Weight training, free weights	281	351	421
Weight training, Hydrafitness	431	539	647
Weight training, Nautilus	300	376	451
Weight training, Universal	379	473	568
Whitewater rafting, kayaking, canoeing	272	340	408
Windsurfing	163	204	245
Wrestling (One match = 5 minutes)	327	408	490

PART FOUR

Diet Analysis Worksheets

The purpose of this exercise is to assess your overall eating habits and compare your nutrient intake to the recommended dietary allowances. This assessment will provide you with the knowledge you need to make positive changes in your diet for overall health improvement.

Directions:

For a 3-day period (two weekdays and one weekend day), eat the typical foods that constitute your normal diet. At the end of each day, record on the daily nutrient intake worksheet the foods eaten for the day and the amount of the listed nutrients contained in each. See the nutritional contents of common foods table and the fast food tables provided in this pocket guide for the listings of the nutrients contained in various foods. Or check the package of the food item as most foods now have the amounts of the nutrients listed on the package. Total the values for each nutrient at the bottom of the chart.

Transfer the total for that day to the nutrient summary worksheet. At the end of the 3-day period, total the daily values and divide by 3 to get the average dietary intake for each of the nutrients analyzed. Compare your average intake for each of the nutrients with those recommended in the following tables for your sex and age group.

Remember that this analysis is only as representative of your normal diet as the foods you eat over the 3-day period.

We have provided additional worksheets for a 3 month and 6 month check-up, so that you can check your progress.

DAILY NUTRIENT INTAKE

Date: ____________________

Foods	Amount	kcals (totals)	kcals from fat	Protein (gm)	CHO (gm)	Fiber (gm)	Fat (gm)	Sat. Fat (gm)	Chol (mg)
Totals									

DAILY NUTRIENT INTAKE

Date: ________________

Foods	Amount	kcals (totals)	kcals from fat	Protein (gm)	CHO (gm)	Fiber (gm)	Fat (gm)	Sat. Fat (gm)	Chol (mg)
Totals									

DAILY NUTRIENT INTAKE

Date: ________________

Foods	Amount	kcals (totals)	kcals from fat	Protein (gm)	CHO (gm)	Fiber (gm)	Fat (gm)	Sat. Fat (gm)	Chol (mg)
Totals									

Three day Nutrient summary

DAY	kcals (totals)	kcals from fat	Protein (gm)	CHO (gm)	Fiber (gm)	Fat (gm)	Sat. Fat (gm)	Chol (mg)
One								
Two								
Three								
TOTALS								
AVERAGE								

TABLE ONE Recommended Dietary Allowances for Comparison

Men 15–18	See Below	< 30 %	See Below	> 58%	~ 30	< 30%	< 10%	<300
Men 19–24		< 30 %		> 58%	~ 30	< 30%	< 10%	<300
Men 25–50		< 30 %		> 58%	~ 30	< 30%	< 10%	<300
Men 51+		< 30 %		> 58%	~ 30	< 30%	< 10%	<300
Women 15–18		< 30 %		> 58%	~ 30	< 30%	< 10%	<300
Women 19–24		< 30 %		> 58%	~ 30	< 30%	< 10%	<300
Women 25–50		< 30 %		> 58%	~ 30	< 30%	< 10%	<300
Women 51+		< 30 %		> 58%	~ 30	< 30%	< 10%	<300
Pregnant		< 30 %		> 58%	~ 30	< 30%	< 10%	<300
Lactating		< 30 %		> 58%	~ 30	< 30%	< 10%	<300

See Table Two for determination of kcal requirements.
Protein intake should be 0.8 grams/kg of body weight (0.36 grams per lb.). Pregnant women should add 15 g. and lactating women should add 20g.
All recommended allowances for macronutrients come from nutrition experts.
Reprinted with permission from "Recommended Dietary Allowances," 10th edition, 1989, by the National Academy of Sciences. Published by National Academy Press, Washington , D.C.

TABLE TWO Estimation of Daily Caloric Expenditure on Body Weight and Physical Activity

To compute your estimated daily caloric expenditure, multiply your body weight in pounds by the calories per pound that corresponds to your activity level.

Activity Level	Description	Calories per Pound of Body Weight Expended during a 24-Hour Period
1	Very sedentary (restricted movement, such as a patient confined to a house)	13
2	Sedentary (most U.S. citizens, doing light work or office job)	14
3	Moderate activity (many college students, performing some daily activity and weekend recreation)	15
4	Very physically active (vigorous activity at least 3-4 times/week)	16
5	Competitive athlete (daily activity in high-energy sport)	17–18

DAILY NUTRIENT INTAKE

Date: ____________________

Foods	Amount	kcals (totals)	kcals from fat	Protein (gm)	CHO (gm)	Fiber (gm)	Fat (gm)	Sat. Fat (gm)	Chol (mg)
Totals									

DAILY NUTRIENT INTAKE

Date: ____________________

Foods	Amount	kcals (totals)	kcals from fat	Protein (gm)	CHO (gm)	Fiber (gm)	Fat (gm)	Sat. Fat (gm)	Chol (mg)
Totals									

DAILY NUTRIENT INTAKE

Date: ________________

Foods	Amount	kcals (totals)	kcals from fat	Protein (gm)	CHO (gm)	Fiber (gm)	Fat (gm)	Sat. Fat (gm)	Chol (mg)
Totals									

DAILY NUTRIENT INTAKE

Date: ______________

Foods	Amount	kcals (totals)	kcals from fat	Protein (gm)	CHO (gm)	Fiber (gm)	Fat (gm)	Sat. Fat (gm)	Chol (mg)
Totals									

DAILY NUTRIENT INTAKE

Date: ____________________

Foods	Amount	kcals (totals)	kcals from fat	Protein (gm)	CHO (gm)	Fiber (gm)	Fat (gm)	Sat. Fat (gm)	Chol (mg)
Totals									

DAILY NUTRIENT INTAKE

Date: ________________

Foods	Amount	kcals (totals)	kcals from fat	Protein (gm)	CHO (gm)	Fiber (gm)	Fat (gm)	Sat. Fat (gm)	Chol (mg)
Totals									

3 MONTH DIET ANALYSIS

Date: ________________

Three day Nutrient summary

DAY	kcals (totals)	kcals from fat	Protein (gm)	CHO (gm)	Fiber (gm)	Fat (gm)	Sat. Fat (gm)	Chol (mg)
One								
Two								
Three								
TOTALS								
AVERAGE								

6 MONTH DIET ANALYSIS

Date: ________________

Three day Nutrient summary

DAY	kcals (totals)	kcals from fat	Protein (gm)	CHO (gm)	Fiber (gm)	Fat (gm)	Sat. Fat (gm)	Chol (mg)
One								
Two								
Three								
TOTALS								
AVERAGE								

PART FIVE

Sample Diet Forms

Once you have completed the diet analysis worksheets, you should have a general idea of how your diet may need modification. Based on your current eating habits and nutrient intake, you may wish to construct a new diet that meets the recommended dietary goals provided in tables one and two. Use the following forms and the information you obtained in the diet analysis exercise to construct a diet that works for you.

	Kcal	Fat (g)	Sat. Fat (g)	Chol (mg)	Fiber (g)	Carb (g)	Protein (g)
Breakfast							
Lunch							
Dinner							
TOTALS							
RDA	1670	< 30%	< 10%	< 300	~ 30	> 58%	40
% of RDA							

	Kcal	Fat (g)	Sat. Fat (g)	Chol (mg)	Fiber (g)	Carb (g)	Protein (g)
Breakfast							
Lunch							
Dinner							
TOTALS							
RDA	1670	< 30%	< 10%	< 300	~ 30	> 58%	40
% of RDA							

	Kcal	Fat (g)	Sat. Fat (g)	Chol (mg)	Fiber (g)	Carb (g)	Protein (g)
Breakfast							
Lunch							
Dinner							
TOTALS							
RDA	1670	< 30%	< 10%	< 300	~ 30	> 58%	40
% of RDA							

	Kcal	Fat (g)	Sat. Fat (g)	Chol (mg)	Fiber (g)	Carb (g)	Protein (g)
Breakfast							
Lunch							
Dinner							
TOTALS							
RDA	1670	< 30%	< 10%	< 300	~ 30	> 58%	40
% of RDA							